© 2019

BEAUTY, METAPHORS

& SO MUCH MORE

Explorations in Science Education

Edited by Punya Mishra

With contributions from: Nicole Bowers, Kevin Close,
Day Greenberg, Bryan Henderson, Sarah Keenan-Lechel,
K.K.Mashood, Rohit Mehta & Paul Reimer

Last Revised: July 2019

ABOUT THIS BOOK

The chapters in this book were previously published in *iWonder: Rediscovering School Science*, and have been included here with permission. The articles (and their titles) have been reformatted and in certain circumstances have been lightly edited for inclusion in the book. Complete citations for the original articles: including author names, titles, date of publication are provided at the beginning of each chapter. All the illustrations for the articles were created by Punya Mishra, unless indicated otherwise.

Current and back issues of *iWonder* can be found at:

https://azimpremjiuniversity.edu.in/SitePages/resources-iwonder.aspx

For questions about this book contact:

    Punya Mishra

    Email: punyamishra@gmail.com

    Web: punyamishra.com

For questions about iWonder, contact:

    Email: iwonder@apu.edu.in

CONTENTS

*Men are mortal. So are ideas. An idea
needs propagation as much as a plant needs
watering. Otherwise both will wither and die.*
~ B. R. Ambedkar

INTRODUCTION

A couple of years ago, during a visit to Bangalore I met Ram G. Vallath. Ram G., as he is known, is a smart, funny guy, a former engineer, like myself, with a deep commitment to science and education. He was, at that time, in the early stages of launching a new science magazine for middle-school teachers. This magazine, which he was co-editing with Chitra Ravi (who I would meet later) would be called *iWonder: Rediscovering School Science,* and was being supported and funded by the *Azim Premji Foundation.* The goal of *iWonder* was

> *...to engage teachers (as well as parents, researchers and other interested adults) in a gentle, and hopefully reflective, dialogue about the many dimensions of teaching and learning of science in class and outside it.*

I was quite excited to learn this. I had grown up on a diet of magazines like *Scientific American* and *Science Today*; my love of science and mathematics built on reading authors such as Stephen Jay Gould, Carl Sagan and Richard Dawkins. I had always dreamed of writing about science for a wider audience, and I sensed an opportunity to make that dream come true.

I also knew that the *Azim Premji Foundation* would put out a quality product. *iWonder* was following on the footsteps of *At Right Angles*, a mathematics education journal they had been publishing for a while, and one that I had already written for. (I had published a math poem as well as a series on mathematics and art with my friend Gaurav Bhatnagar).

So I pitched an idea to Ram G., that of writing a series bringing what we have learned from science-education research to the classroom. With a bit of discussion we had a plan and this series called *Research to Practice* was born.

Once we had a few of these articles in place, it seemed like a good idea to bring them together as an e-book, hopefully to garner a wider audience.

All these articles are co-authored with either one or more graduate students. This is an intensely collaborative process, that starts from choosing the topic, reviewing the research and then working together to craft an accessible and interesting article. Academic writing can be stuffy, laden with jargon that gets in the way of message. Writing for practitioners means that we have to focus on clarity of expression without undermining the complexity of the ideas. Whether or not we succeed is for you, the reader, to determine, but I do believe this is an important goal to aim for.

There has been an additional, unanticipated, benefit to writing these articles. When we were writing the first article, I decided, just for the heck of it, to create a set of illustrations that would go with the article, and, out of their generosity of heart, Ram G. and Chitra were kind enough include my illustrations in the final article. Thus began a new sub-career as a scientific illustrator, and I have illustrated all the subsequent articles. What fun.

A project of this nature would not be possible without the

help of a bunch of people. First and foremost Ram G. and Chitra for giving me (and the students) the opportunity to write these articles. Thanks also to all my co-authors for their passion and commitment to this work. We owe a special debt of gratitude to the anonymous reviewers for their feedback. And finally, a special shout out to Rohit Mehta, who was an author on the first piece we published, and is now moved on to become a co-editor in this series.

*All my life through, the new sights of Nature
made me rejoice like a child*

~ Marie Curie

WHY CARE ABOUT BEAUTY IN SCIENCE TEACHING

Horses and rainbows make the world seem more exciting, not science—student quoted in Mark Girod's dissertation research study

He to whom the emotion is a stranger, who can no longer pause to wonder and stand wrapped in awe, is as good as dead —his eyes are closed. – Albert Einstein

Mehta, R., & Keenan, S. (2016, June). Why teachers should care about beauty in science education. *iWonder: Rediscovering School Science (1) 2*, 83-86.

Scientists often speak of being inspired by wonder and a sense of beauty. They describe the universe with a sense of awe about it and our place in it, the drama in searching for the truth, and the elegance of scientific ideas and structures. Consider this quote by Richard Feynman:

> *The world looks so different after learning science. For example, trees are made of air, primarily. When they are burned, they go back to air, and in the flaming heat is released the flaming heat of the sun which was bound in to convert the air into tree, and in the ash is the small remnant of the part which did not come from air that came from the solid earth, instead. These are beautiful things, and the content of science is wonderfully full of them. They are very inspiring, and they can be used to inspire others.*

Feynman highlights the potential of wonder and inspiration as he contemplates the world around him and the intricacies of its inner workings. This sense of wonder opens new doors and further questions to pursue; it is the beginning of a curiosity that inspires more questions. Science, he suggests, helps us find answers to some of the most awe-inspiring questions we have about nature. It is not a rigid method, or stuffed with facts and information that we need to memorize to pass an exam; but, rather, a rich and exciting process, an adventure, and a journey to solve the mysteries of the world.

This sense of engagement and passion is often in stark contrast to how many students in school think about science. Science is often seen as being full of arbitrary facts, mindless activities and, thus, quite dull and boring (as seen in the quote at the beginning of this article). To be clear, we are not suggesting that scientific facts and theories are not important. Neither do we seek to imply that the scientific method is not rigorous and demanding, or, not in itself a reason that motivates some scientists. Our goal is to point out that what

Science is one of the most powerful ways to engage with the beauty of the universe. We use science to understand the cosmos and, in the process, find beauty in our understandings and representations of it.

motivates scientists is not just the facts or the method per se, or instrumental reasons (such as economic viability), but also their passions that emerge from the excitement of the chase; the beauty, elegance and explanatory power of scientific ideas. The quote by Einstein, that opened this article, suggests that science is not a dispassionate activity. It is as much about facts as it is about wonder, passion, emotion, and beauty. This is what we call an aesthetic perspective.

What Does the Research Say?
What can we do to make science come alive for our students? How can we get them to appreciate the beauty and wonder of scientific ideas? What if we took some of these aesthetic elements that scientists speak of and bring them to the forefront of teaching science? What would happen then? Would students respond differently to what they were learning? Would their ideas of science, and what it means to do science, change?

One educational researcher, Mark Girod tried to find answers to these questions. He argued that the aesthetic experience is not just restricted to the arts, but is an integral part of doing and learning science as well. He suggested that by building on the emotional and affective elements of doing science, we could motivate our students to wonder more deeply about nature and science, stimulate their curiosity and interest, and thus transform their experience of learning science.

In his research, Mark studied two 4th grade science teachers. One of these teachers, Ms. Parker, was an experienced and accomplished teacher who taught science in the traditional way, focusing on facts and conceptual understanding. The other teacher, Mr. Smith, was also an experienced teacher, but held a different focus. Mr. Smith's class was designed to foster excitement and interest by organizing content around the power ideas have to inspire and renew perception, providing opportunities for students to experience the world in new ways that consistently highlighted the aesthetic and artful aspects of science. For example, conducting a class in a garden, while admiring the beauty of flowers, and framing questions that help explain, say, where a flower gets its color.

Mark's research showed that at the end of the day, students in Mr. Smith's class not only performed better than the students in Ms. Parker's class on standardized tests, but also showed greater engagement with scientific ideas speaking about how they had discussed these ideas with their family and friends outside of the classroom. In short, students in Mr. Smith's classroom were drawn to wonder, inspired to discuss it with others, and enjoyed seeing the world through the lens of scientific ideas.

How Can Teachers Cultivate an Aesthetic Classroom?

So what did Mr. Smith do in his classroom? While interested teachers may want to read Mark's study, we offer three suggestions that Mr. Smith successfully used in his classroom.

Guide 1: Frame content around metaphors and perceptual lenses
While covering the topic of weather and the atmosphere, Mr. Smith did not just describe terminology and facts—he made them real through the use of powerful metaphors. For instance, he had his students lie down on the grass and look up at the sky – and told them about the ocean of air, 17 miles deep, pressing down on them. In other words, he shared powerful facts with his students: ideas of ways of looking at the world in an effort to generate a sense of wonderment. Lying on the grass looking at miles of vast sky and thinking

of a metaphoric ocean stirred an aesthetic chord with Mr. Smith's students that a simple lecture could not.

How to implement: When thinking of a metaphor, make sure that you come with an idea that works on similar physical principles. It makes your work as a teacher more powerful when the transferability of the laws of physics makes the experience meaningful for your students. In this example, because air and water are both fluids, the metaphor makes conceptual sense and helps students understand and remember the ideas.

Guide 2: Making it personal and learner-oriented
Mr. Smith constantly tried to empower his student to see and act with science in ways that fit them individually. He had them share science related stories from their lives prompted by questions such as "who thought about wind yesterday? What did you think about?" He pushed them also to "re-see" the world in new ways based on their learning of science. He also modeled for students how he saw the world through the lens of science–purposefully using words that demonstrated a connection between art and beauty and science. Even when using traditional worksheets, Mr. Smith included at least one question that allowed students to comment on their personal experiences with science content. Here, an aesthetic motivation can expand and enhance student experiences in science.

How to implement: A critical thing to consider is knowing what your students care about – getting to know their personal preferences and interests opens a window into their lives. You can then highlight those aspects of science that intersect with their lived experiences, in ways that they may have never thought of before. This makes the concept come alive in their minds, enriching daily life.

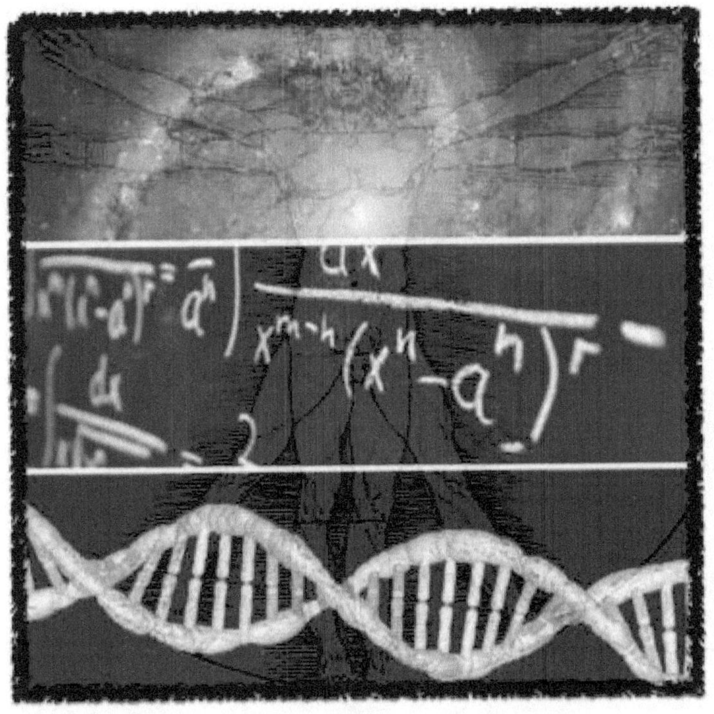

Fig. 2. Connecting across scales of beauty. From the grandeur of the cosmos to the intricacies of sub-atomic particles, beauty is all around us. These infinities (of the very small and the very large) are bridged by the human intellect—the beauty of mathematics at work.

Guide 3: Developing group activities that emphasize the aesthetic experience of learning science

Mr. Smith constructed a range of activities in his classroom designed to facilitate emerging aesthetic understanding and new ways of seeing the world. Students were asked to look at how artists use the sky to convey emotion, or to actually have them create artwork that attempted to capture similar ideas. He had them undertake "field-trips"—short walks around school to observe science ideas learned in class. He had them make models from gumdrops and toothpicks as well as play with make-believe scenarios about upcoming weather events.

How to implement: Scientific ideas are powerful but, often, alien to us. Consider the vastness of space and the scale of the solar system. Making this scale realistic gives students an opportunity to aesthetically experience and feel a sense of awe and wonder. Let your students collaborate to embody a solar system in your playground to calculate the relative distance of the planets from the Sun. Come up with a scale for distance and let one student be the sun, while other students become the planets. How far apart would they be? Is it even possible to go as far as Neptune or Pluto without leaving the school property? If one student becomes a beam of light, how long does it take to go from the Sun all the way to Neptune? And, then, imagine the next closest star being 4 light years away! Have your students attempt to visually and artistically represent these distances and scales.

Conclusion

If science lets us perceive the world around us in new, transformative ways, then our job as teachers is to facilitate that re-visioning process. In fact, there is research to show that a teacher's beliefs about the nature of science are an important factor in how science is perceived by students. We need to step away from rigid curricula that focus only on success in tests, and, instead bring to our classrooms a sense of wonder and appreciation for the beauty of scientific ideas. We hope that the three broad suggestions above are just the beginning of many different ways in which science ideas can come alive in the minds and lives of our students.

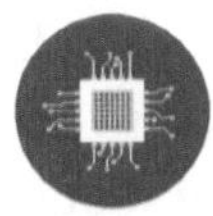

SOCIAL JUSTICE AND SCIENCE EDUCATION

There is a need to prepare teachers and students for the new roles that they must play... Our vision of schooling, and by extension science education, is more aligned with participatory democracy where citizens actively collaborate... for future generations. —
Mike Mueller

I used to probably think, well I thought that, um, school wasn't that important, but then this program made me feel that I belonged in the school, that I made a change, so I had to go to school, cause it helped me and it helped the school, a lot. —
Fatima, an elementary student who did "action research" in school

---

Greenberg, D. (2017, March). Why science teachers should care about social justice. *iWonder: Rediscovering School Science (1)* 1, 70-73.

As our world continues to change along multiple dimensions—socially, politically, ecologically, medically, digitally, and economically—it is increasingly important for science teachers to provide students with the skills, knowledge, and opportunities to apply their learning to the world around them. They should seek to empower their students with the agency and tools to confidently meet challenges of the present and the future. Far too often, however, teachers assume that their students are not capable of accomplishing big things as young people. This is especially true for students who come from communities that have been excluded from mainstream science learning and careers (e.g., students who are members of immigrant or ethnic minority communities, and students from low-resource schools and neighborhoods).

What is Social Justice Science Education?

Science education for social justice involves using science knowledge and skills to question existing systems of power - systems that oppress students and their communities. Science teachers who work for social justice use their classroom practice as a tool of social, political, and academic empowerment for their students. They make spaces for their students to make change for themselves and their world, in partnership with others. These teachers empower students by making their science classroom a place where students can build skills and use knowledge in ways that mirror and/or speak back to what the real world already asks of them (and will continue to ask of them in the future). They teach their students how to navigate issues and challenges that involve science knowledge and practices, critically question/examine related factors, and ultimately act for positive change and greater justice using science as a tool for change.

Science and social justice – finding the balance!

Science teachers can empower their students (socially, politically, and academically) to use science in ways aimed at making the world a fairer and more just place for everyone. Integrating social justice into their classroom practice offers teachers the tools they need to accomplish this goal.

Social justice oriented science education involves creating learning spaces where students build skills and use knowledge in ways that mirror and/or speak back to what the real world already asks of them (and will continue to ask of them in the future). Students learn to question existing systems of power, systems that oppress students and their communities. Science teachers can show their students how to navigate socio-scientific issues and challenges – issues that involve science

knowledge and practices along with a critical questioning/
examination of related social factors. Ultimately, students
can learn to use science as a tool to act for positive change
and greater justice, in partnership with others. This can
benefit science learning, because the shared desire to make
a specific positive change as a class can motivate students
to reach related science learning goals. Then, students can
activate their ownership of content (not just absorption) to
take action themselves and/or convince others in positions
of power to create positive change. For example, students
who live in a flood zone could work towards physics learning
goals to inform a student-organized public communications
campaign to educate local community members on how to
design stronger flood control structures. In this example,
relevant physics content (e.g., energy, force, speed, weight,
etc.), can become more directly important and meaningful
when students recognize it as a tool to inform and protect
their families and friends.

What does the research say?

How can science teachers make science matter for their
students? How can they provide students with the tools to
become leaders who take educated and meaningful action
with science?

Educational researcher Alexandra Schindel Dimick studied
a case of high school environmental science teaching that
was organized around the goal of community action for
greater social justice. Her study focused on science teacher
Mr. Carson, who introduced social justice actions into his
high school environmental science curriculum. He discussed
local environmental problems with students, and allowed
them to form three separate working groups to address these

problems. His students learned about pollutants in their local waterway and conducted lab experiments related to water chemistry. They then created posters to educate the public on science topics related to water. Finally, they completed group action projects to solve problems in their local waterfront environment.

In her observations on the work done by the students, and her interviews with some of them (9 from a total of 24 students in the class), Alexandra found that students felt an increased sense of power when their learning was directed towards social justice. Mr. Carson explained:

> ... [the students] are feeling empowered to change something that affects them and they're not depending on other people to make the change. They're the ones who are being affected; they're the ones that are trying to make the change.

However, it should be noted that not all of Mr. Carson's students felt that the project was a complete success. Social justice science education must empower students socially, politically, and academically. Alexandra's study showed that the students noticed when one of these dimensions was missing, and they shared their disappointment in not receiving the support they needed to succeed as action-takers in science. It should also be noted that although Mr. Carson taught environmental science, social justice action can be integrated into every type of science classroom.

How Can Teachers Become Social Justice Science Teachers?

What can we learn from this study about the successes and failures of implementing ideas of social justice in science classrooms? Here are three components that appear to be important to enact in order to successfully teach science for social justice:.

Participatory engagement in a classroom context. In the background is a word cloud created from all the words in the Wikipedia page on "Science."

Guide 1: Support Student Leadership and Collaboration (The Social Component)

Mr. Carson had students divide themselves up into three separate groups after they voted to narrow down their ideas for taking action. He did not, however, provide group members with any specific strategies or guidelines for making group decisions and sharing discussion time. As a result, group dynamics affected individual actions, decreasing speed and efficiency.

How to implement: Allow students to choose their own working groups based on topic, but provide them with tools to navigate group dynamics. Give them opportunities to practice healthy teamwork and compromise, within a supportive structure of group social rules that students can agree on together. Have students create a social contract together that dictates how to behave in groups and how to resolve conflicts. Empower your students to work well together by helping them create shared expectations and the tools to solve problems themselves.

Guide 2: Prepare Students to Take Action (The Political Component)
Mr. Carson removed himself from classroom interactions after students formed groups. His students lacked the support they needed to transition from understanding action in terms of daily, personal responsibilities to understanding action in terms of larger, more transformative, active participation in change. As a result, some students came to a larger understanding of action with science themselves, but other students ended the project without knowing how to take further, larger, or more complex action for social justice with science.

How to implement: Be an active part of student interactions to make changes that matter to them and their community. Discuss the difference between individual action and collective action, and help students to critically question structures of power that cause pain to people or destruction to environmental systems. Help them discover the "root causes of problems" that often relate to social injustice but can also be better understood through scientific investigation. For example, science teachers could support their students in asking: How does a local school building's mold infestation relate to political debates on education budget proposals? How has genetics been used to marginalize existing populations? Who profits from the use of coal energy and

The dove with an olive branch, a symbol of peace. Made from icons representative of science, flies against a word-cloud created from all the words in the Wikipedia page on "Social Justice."

mining? How have the contributions of women to computer science been neglected in histories of computing?

Guide 3: Help Students Use Knowledge as a Tool (The Academic Component)
Student empowerment can align with teaching and learning science education when social justice is the goal and science knowledge and practice is the tool. In order for students to feel empowered with scientific knowledge, teachers must actively facilitate experiences and provide supportive resources and information. It is not enough to give students the opportunity to lead—science teachers must also arm students with the scientific skills and knowledge to take the action that they desire.

How to implement: Each component of this model is important, but teachers must use all three to support social justice actions with science knowledge. Help students to learn about the ways in which actions they want to take can be supported with science learning and practice. Once they have a big goal in mind, help students break that down into smaller, accessible goals and have them lead a discussion together about what information and skills they need to learn in order to reach each smaller goal. As they achieve the small goals, continually check in with them to discuss how their step-by-step progress is bringing them closer to their larger goal of big change (at the same time, help them see and feel proud about how their collection of knowledge and skills is growing in size and depth).

Conclusion

Science teachers must offer their students the resources and support to combine social, political, and academic growth in order to take action for social justice with science. When Mr. Carson did not combine and equally address all three components, his students were disappointed. They described their socially and politically informed actions as not being supported enough with academic knowledge and connections to their classroom learning. When, however, all three aspects work together, the results are positive. As Janis, a 13-year-old student in another social justice science teaching project explained:

> *We know what we are doing. We know how to make a difference. We know how to save energy and how to convince other people of better ways to do things with electricity. That is one way that we are experts.*

Teachers should empower their students with science, in ways that provide students with the tools to solve real problems that affect them and the people and surroundings they love and depend on. This requires more than simply giving students permission to be powerful actors. It requires supporting students as they work to complete each step along the way towards addressing bigger and more complex issues, with scientific, social, and political knowledge and tools.

SEEING THE WORLD THROUGH MULTIPLE METAPHORS

'Our ordinary conceptual system, in terms of which we both think and act, is fundamentally metaphorical in nature'.

– George Lakoff.

'To see a world in a grain of sand

And a heaven in a wild flower

Hold infinity in the palm of your hand

And eternity in an hour'.

– William Blake

Mashood, K. K., Mehta, R., & Mishra, P. (2018). To see a world: Using multiple metaphors in science education. *iWonder : Rediscovering School Science* (1) p. 48-52.

How do we learn something new? It can be argued that the only way we can understand something new is in terms of what we already know. This idea lies at the heart of metaphors. This is of significance when we seek to understand abstract ideas. For instance, consider the idea of energy, a complex foundational concept that is discussed seemingly differently in different domains of science. The challenge here is to connect the abstract idea (in this case, of energy) to something that learners already know, yet to do it in a way that prevents unitary, simplistic understandings. In this article, we consider the example of teaching about energy through multiple metaphors to provide students with a rich understanding of the concept.

As educators know, teaching a human being is not like programming a robot or a computer to do something. In the latter case, all we need to do is provide accurate instructional inputs. There is no room for interpretation or imagination, or for alternative conceptions or misconceptions. In other words, what is 'told' is 'taken' as it is, provided there are no technical glitches or errors. But, teaching a human being is complicated, even when we do it with—what we think is—absolute clarity. Consider the following remark from Carl Wieman, Nobel-winning physicist turned education researcher (Wieman, 2007):

> *"When I first taught physics as a young assistant professor, I used the approach that is all too common when someone is called upon to teach something. First, I thought very hard about the topic and got it clear in my own mind. Then I explained it to my students so that they would understand it with the same clarity I had. At least that was the theory ... [But] whenever I made any serious attempt to determine what my students were learning, it was clear that this approach just didn't work. An occasional student here and there might have understood my beautifully*

Why does this happen? It is unlikely that someone like Wieman does not know the physics he is trying to teach or lacks clarity in its exposition. The real reason could be the underlying complexity of human cognition. We do not necessarily learn anything new exactly as it is told to us, but try to comprehend it using what we already know. Learners are active constructors of knowledge in which a range of factors (such as their prior experiences, social context, linguistic ability, and emotional setup) can influence how they receive, store and retrieve information.

The fact that learning builds on, is constrained, and framed by, what we already know is often taken to be a problem, since it widens the factors that educators need to consider as we design our lessons. However, we argue that as educators we need to accept this and learn to harness it to our advantage. One strategy for building on prior knowledge is to use metaphors to explain new ideas. In fact, metaphors are something that all teachers have knowingly or unknowingly used at some point in their work.

We need to be sensitive, however, to the limitations of using single metaphors to explain a complex idea. Single metaphors can cause a seductive reduction in complexity of an otherwise rich idea (See Fig.1). In this article, we focus on one such strategy—of using multiple metaphors to explain complex scientific ideas. We ground our discussion of this idea in one example—that of teaching about energy.

Teaching the Concept of Energy

Energy is one of the most fundamental and overarching concepts in science. Students learn about energy in school in

The seduction of using single metaphors: using just one representation reduces the complexity of a rich idea. Illustration inspired by Variante de la tristesse by Belgian surrealist René Magritte

a variety of contexts. In biology, the idea of energy is the key to understanding important topics such as photosynthesis and nutrition. Energy in chemistry is integral to understanding concepts such as chemical bonds and chemical energy. In physics, energy is discussed in terms of work, in forms of kinetic and potential energy. Often the concept of energy – irrespective of what aspect of science is taught – is introduced by giving textbook cliché definitions like "energy is the ability to do work."

We reason that such an approach hardly encompasses or conveys the underlying conceptual richness of this important and complex concept. In addition, students hear the same word in different classes/contexts and are not provided opportunities to connect across these seemingly different uses of the same word. Thus, students can hold an incomplete and fragmented understanding of this concept. And, if these misunderstandings are related to core science concepts (such as energy), it can lead to further confusion and misconceptions down the road.

Multiple Metaphors of Energy: The Research Perspective

Rachael Lancor's research provides a perspective into the range of concepts that underlie the concept of energy and thus provide insights on how the concept of energy can be taught effectively (Lancor, 2013, 2014). She considers how physics, chemistry, and biology textbooks (along with science education literature) conceptualize energy, and through this gives a picture of the web of ideas essential to understanding the concept. The study finds that each discipline, on its own, focuses only on a subset of the overarching concept. It is only by providing a framework that integrates these seemingly distinct views that we can hope that students truly understand it.

Lancor begins by stating five main characteristics associated with the concept of energy that students need to understand. These, relatively abstract ideas, are:

- Energy is conserved.
- Energy may dissipate from a system over time.
- Energy can be transformed from one form to another.
- Energy is transferred between parts of a system, and that
- Energy could be received by a system from another source.

Lancor then follows this up with an analysis of introductory science textbooks (across domains such as physics, biology, and chemistry) to see how these abstract ideas are explained. In brief, she identifies 6 different key conceptual metaphors that ground the abstract ideas in ways that make sense to students. These are that:

- Energy like money, can be accounted and tracked.
- Energy can take different forms, and changes from one to another.
- Energy can flow like water through a pipe.
- Energy can be carried by organisms, as well as inanimate entities like electrons.
- Energy like oil in a faulty machine, can be lost.
- Energy can be stored in devices like a battery or a wound spring.

Of course, each of these metaphors, if used in isolation, leads to the distinct possibility that students would develop a limited understanding of the nature of energy, and how it plays out across different scientific disciplines. It is here that the idea of weaving multiple metaphors together can help convey both the richness of the concept and prevent students from holding on to one metaphor as being the only way to understand a concept (see Fig.2).

What is important to note is that each of the six metaphors of energy highlight only some of the five main characteristics of energy and obscure others. For instance, treating energy as a substance that can be tracked and accounted highlights the aspect that energy is conserved but obscures the characteristic that it can be transformed from one form to another. On the other hand, the metaphor of energy as a substance that can be lost highlights the fact that it can dissipate but obscures the idea of conservation of energy. So, a coherent understanding of energy would require careful considerations of the various characteristics associated with the concept. Picking a few would result in a situation like that of the blind men trying to describe the elephant—each one understands only a part and can never agree overall. This necessitates weaving multiple metaphors together to reach a richer understanding of the whole picture.

The six metaphors for energy. Illustration inspired by the painting 'The key to dreams' by Belgian surrealist René Magritte.

Energy is often introduced in school classrooms by giving a short definition, like "energy is the ability to do work" (Lancor, 2013, 2014). This is not only ineffective but provides students with an incomplete picture about a rich, critically important, foundational concept in science. Students merely memorize such clichéd statements and regurgitate them during examinations. This also has the implicit danger of giving students as well as teachers a false sense of knowing. A discussion of energy based on multiple metaphors may help resolve some of these issues. In addition to deepening student understanding, it can also facilitate connecting ideas and disciplines. Below we discuss some guidelines on how to implement them in science classrooms.

Learning to think like scientists.
Unlike scientists and experts, students are less likely to be proficient with the nuances associated with language. They may not have the depth of understanding to precisely interpret metaphors and may take things literally. What we often call as misconceptions or alternative conceptions pertaining to energy are mostly imprecise re-conceptualizations of metaphors. The expertise and deeper understanding to appreciate these metaphors cannot be obtained overnight. It requires continuous engagement and development over time. It is, therefore, important to demonstrate the iterative process of scientific method.

How to implement? Feynman's lectures on physics provide an example of building arguments pertaining to energy conservation (Feynman et al., 2015). He starts with the story of the fictional character of Dennis the Menace playing with a set of blocks. The story containing metaphorical descriptions keeps growing in complexity with each block, with mathematical equations making their entry towards the end. Similarly, there should be considerable engagement with the concept of energy in terms of metaphors. Once it is ensured that students have a considerable grasp with the metaphorical language and discourse pertaining to energy, the teacher may bring in formal definitions, equations, and numerical problems. Encouraging students to explore connections between non-mathematical and mathematical concepts will further boost coherence and rigor. This may often require revisiting the metaphors and further strengthening the concepts iteratively. This can be done in small groups.

Transdisciplinary thinking.
Concepts like energy are present across disciplines. However, how the idea of energy is discussed in physics may differ (at least at a surface level) from how energy is discussed in biology or chemistry. A specific topic in a discipline may afford a certain metaphor (Lancour, 2013). A coherent, in-depth understanding of the concept requires reconciliation between its apparently different disciplinary manifestations. Unfortunately, this conceptually challenging task is often left wholly to students. A discussion based on multiple metaphors facilitates reconciliation between energy in physics, chemistry, and biology, and has the additional benefit of providing richer language and conceptual structures to talk and reason about these ideas.

How to implement? Identify the metaphors associated with a specific description of energy. For example, biology discusses the flow of energy through ecosystems. The metaphor of energy as a substance that flows like water through a pipe is relevant to this context. On the other hand, discussion of a stone rolling down a hill is found in physics. Here the metaphor of energy as a substance that changes from one form to another is apt as potential energy is getting

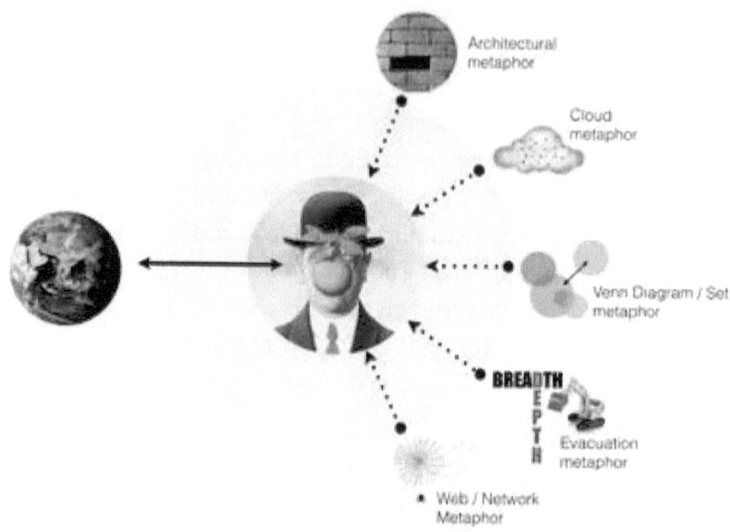

Multiple metaphors of knowledge in understanding the world. Illustration inspired by paintings by Belgian surrealist René Magritte

converted to kinetic energy. Highlight the fact that though the metaphors are different, there is an underlying connection between them. It is only when we hold them together coherently that we achieve a complete picture of energy as a concept. Talk to other science teachers in your school and prepare complementary metaphors for physics, chemistry, and biology.

Conclusions

Our understanding of the idea of knowledge itself is metaphorical in nature. For instance, viewing knowledge as a web or network has very different consequences from having an architectural metaphor. The figure above shows some of the metaphors that philosophers have identified for how we conceptualize knowledge. Going deeper into these different metaphors is beyond the scope of this article but something that all educators need to keep in mind.

To be able to see the same thing from different perspectives and fathom the underlying unity between apparently distinct ideas is a wonderful experience. We can help our students climb to that vantage point, where they could appreciate concepts like energy beyond the confines of textbook definitions and equations. Connecting ideas and disciplines is important for multiple reasons (Ledford, 2015). The foundations for transdisciplinary thinking can be built as early as school level. And, as illustrated with the concept of energy, the use of multiple metaphors to teach science has the potential to facilitate such connections between and across disciplines.

Sculpture inspired by Magritte's painting 'The red model"
Photo credit: John Sheehan (@dogstar7tweets)

STUDENTS AS TEACHERS

Learning is least useful when it is private and hidden. It is most powerful when it becomes public and communal ~ Lee Shulman

The facts of science and, à fortiori, its laws are the artificial work of the scientist; science therefore can teach us nothing of the truth; it can only serve us as rule of action ~ Henri Poincaré

Close, K., Bowers, N., Mehta, R., Mishra, P., & J. Bryan Henderson (2018) Students as teachers: How science teachers can collaborate with their students using peer instruction. *iWonder: Rediscovering School Science* (5). p. 24-28.

Teachers often forget that science is social and rhetorical in nature. Group consensus and peer review, not political discourse, define scientific facts. Therefore, scientific instruction should embrace the push-and-pull and back-and-forth of scientific dialogue and argumentation. When instructors speak of the scientific method in school they focus on generating hypotheses and conducting experiments, but often fail to present the whole process as what it really is: a way of crafting a convincing argument. In other words, science is a way of harnessing facts, logic, and evidence in order to convince others that a particular idea is likely correct.

With this in mind, we present a way to introduce meaningful scientific discussion in the classroom by using the most valuable classroom resource, students themselves. This technique, called Peer Instruction, wrests control from the teacher and gives it to students. According to a recent paper by Dr. Trisha Vickrey, a professor in Chemistry at the

Learning is least useful when it is private and hidden, it is most powerful when it becomes public and communal—Lee Shulman.

University of Nebraska-Lincoln, and four of her co-authors, Peer Instruction is one form of research-based instructional reform that has been widely adopted among instructors in science, technology, engineering, and math.

As a widely adopted form of teaching, Peer Instruction allows students time to talk, debate, and teach each other during instruction. Students can serve as tutors, models, and sounding boards for their peers. In fact, research shows that much learning occurs during these peer-to-peer interactions. According to John A.C. Hattie, a renowned professor of education, "if you want to increase student academic achievement, give each student a friend." Social interaction, he asserts, drives students to become their own teachers. It is this social interaction that Peer Instruction seeks to provoke.

Additionally, Peer Instruction replicates something that is fundamental in the scientific process: convincing others of the "truth value" of one's approach. The key aspect here is that this is a process of argumentation—or marshaling data and logic to explain one's point of view. This rhetorical turn is crucial, forcing students to not only come up with the right answer, but also to convince others of the same. A student with a different explanation would approach their partner's statements with a questioning attitude—and also an open-mindedness to being wrong. What is interesting is that such conversations do not necessarily mean that the student with the right answer necessarily believes their own logic. A situation could arise where a student with an incorrect understanding manages to convince their partner (who may have had the right explanation). In fact, this situation reveals the weakness of their understanding, even though they came up with the right answer. Essentially, Peer Instruction stresses conceptual understanding, and the logic of the argument, over merely getting the correct answer.

According to Drs. Eugene Judson and Daiyo Sawada, of Arizona State University and the University of Alberta, respectively, interactive voting in science classrooms, often with flashcards or "clicker" systems, has been around since the 1960s and popular on some campuses since the mid-1990s. One researcher interested in these voting systems, Eric Mazur, a Professor of Physics at Harvard who wrote the book that coined the phrase Peer Instruction, discovered that students, under certain circumstances, learned more if they discussed their answers with their peers after voting. He called this Peer Instruction.

Specifically, Eric Mazur determined a specific model for implementing Peer Instruction:

- Pose a question
- Give students time to think
- Have students record their individual answers
- Have students convince their neighbors (peer discussion)
- Have students record their revised answers
- Calculate the results
- Explain the correct answer

Mazur's findings and his model for Peer Instruction inspired a generation of follow-up research. Further studies by Mazur and his colleague, Catherine Crouch at Harvard university, showed that students who took an introductory physics course with Peer Instruction consistently and significantly outperformed those who took a course without Peer Instruction. To measure performance, Mazur and Crouch gave conceptual physics tests before and after classes for a period of ten years to look for differences between students in a class with Peer Instruction and students in a traditional lecture class. Often, students in the Peer Instruction class

A diagram of the 7 steps in the peer-instruction process

showed learning gains twice as high. Other studies showed that Peer Instruction impacted fields besides physics, such as geoscience, computer science, and calculus, thereby suggesting that Peer Instruction may be suitable as a general teaching strategy, not just confined to physics.

Yet other studies delved into the particular steps needed to implement Peer Instruction effectively. These are lightly modified version of the 7 steps above and include: posing a conceptually challenging question, giving students time to think and give an individual response, having students convince their neighbor to accept their response as correct, having the class record the revised answers, and finally, explaining the correct answer to the class. Though this procedure appears straightforward, each step contains subtle considerations for effective implementation. The following sections will unpack each step and give examples of best practices.

How Can Teachers Support Peer Instruction in Their Classrooms?

Guide 1: Choose the Right Question
Teachers know that questions differ in their degree of challenge. Questions that fit well with Peer Instruction represent a conceptual challenge for the students. A test question like, "Name the phases of mitosis in order," provides students with the opportunity to recall information. A test question like, "How does alternation of generations represent an effective evolutionary turn for the survival of some plant species?" requires students to think through several concepts and link them together. Recall questions do not require Peer Instruction; simply providing the correct answer allows students to understand how their own answer was incorrect. Providing the correct answer for a conceptually challenging question does not allow students to understand how their answer was not correct. Only providing answers honors the answer above the explanation. Without access to and practice with explanation of phenomena, students cannot truly understand concepts deeply or understand the central nature of argumentation in science.

Regarding this central nature of argumentation in science, Peer Instruction strengthens the conceptual fluency of the students with correct answers. Students with correct answers may or may not fully understand all of the concepts behind the correct answer. Peer Instruction provides students a chance to talk through their thinking, particularly when a peer asks questions, thereby helping them think through their response and articulate it so that another student can understand it. Challenging questions require a deeper level of understanding, and Peer Instruction allows students to work through that deeper level of understanding together regardless of their initial answer. Essentially, Peer Instruction emphasizes understanding over merely getting the right answer while allowing students to participate in the authentic practice of argumentation in science.

How to Implement: Think of Peer Instruction as a perfectly timed learning opportunity for your students. Will any old question do? Clearly, factual questions with answers that can be looked up do not work as well. The trick, research shows, is to choose questions that focus on concepts, not on facts. Also choose questions that incite curiosity – questions that may divide the class. Often, questions based on common misconceptions (e.g., "in a frictionless world, which falls faster: a bowling ball or a tennis ball?") drive rich peer discussion.

Guide 2: Elicit Individual Responses
It seems counter-intuitive that individual responses are necessary for a technique called Peer Instruction, but research studies show that Peer Instruction does not work without this crucial step. It is important for students to think initially through the question because it lights the fire of curiosity in the student. They make a decision and commit to it. When students do not engage in this step, peer discussion lacks robust peer critique. In these cases, less confident students often just go along with more confident students without deep discussion. Individual responses allow students time to engage the question thoughtfully once without the influence of the peer. This initial commitment produces a deeper discussion between the peers.

How to Implement: Ask the question and allow students to share their individual answers in class. Introduce paddles, white boards, or flash cards (or, if you have access, technology like

iClickers or free online tools like Braincandy.org) for them to
brainstorm and note their answers. Give them time and space
to form an individual response and commit to it. This creates
more engagement in the peer discussion to follow. Students
report that taking the initial responsibility to answer the
question individually forces them to think more deeply about
the question and the answer.

Guide 3: Peer Discussion
Implementing peer discussion may be the single most
important part of the whole Peer Instruction process.
However, not every question posed to the class requires peer
discussion. Research shows that when a question is too easy
(over 70% of students get the correct answer on the first vote),
teachers should just skip peer discussion because learning
gains are negligible. If the question is too hard (under 35%
of students get the correct answer on the first vote), teachers
should provide more explanation or hints before discussion.

Additionally, teachers should prompt students to discuss not
only their answers, but the reasons behind their answers. This
is key because the focus of learning should be on conceptual
understanding rather than getting the right answer. Research
shows that when teachers prompt "reason-centered"
discussions instead of "answer-centered" discussions learning
gains increase.

How to Implement: Observe students carefully during the
first round of voting to see if the question is too easy or too
difficult for peer discussion. If the question lies in the sweet
spot, then encourage students to turn to their neighbor and
explain why they chose their individual answer. Give the
students time to discuss, trust that they are acting as their own
teachers.

Guide 4: Explaining the Answer
Peer Instruction is incomplete without a final explanation
by the teacher after voting, discussing, and re-voting. Studies
indicate that combining peer discussion with instructor
explanation outperforms other similar pedagogical
approaches. Presumably, this is because students are now

---

*The value (and stages) of developing a shared understanding
through the Peer Instruction Process*

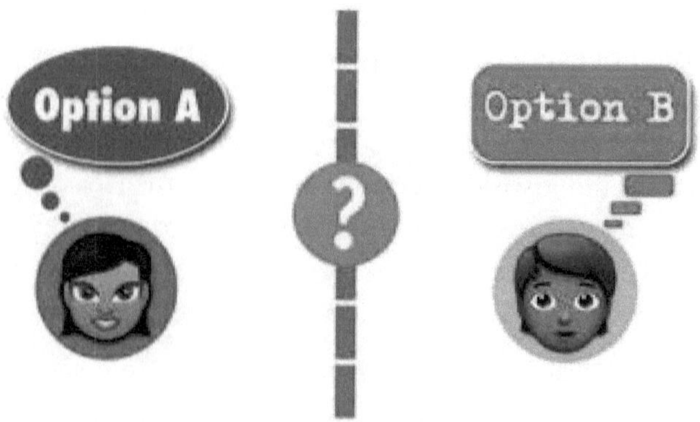

*Students when faced with a problem work in isolation with little knowledge
of each other's understanding.*

*By sharing each other's solutions students realize that their perspectives,
frameworks, and understanding differ from each other.*

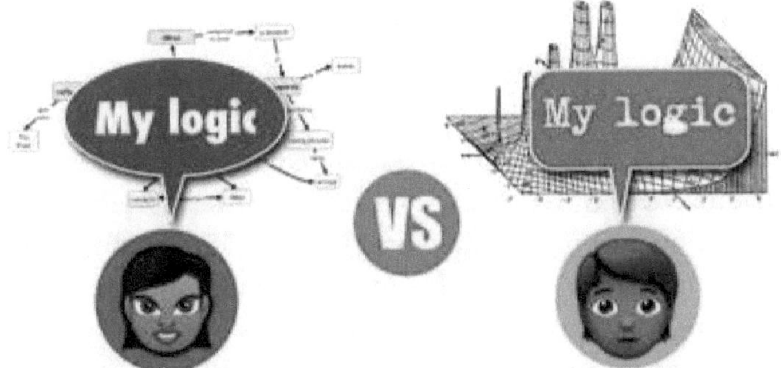

In attempting to convince each other of the correctness of their solution, students have to explain their logic and perspective on attacking the problem at hand.

Through explaining their logic the students have a higher probability of developing a correct shared understanding of the problem.

---

primed and motivated to hear the instructor explanation. Was their first answer (i.e., their individual answer) correct? Or the new answer co-created with a peer? After first steps of the Peer Instruction cycle, students are ready to hear their instructor's point of view.

How to Implement: Once student re-votes are collected, identify and explain the correct answer. Try to draw on some of the popular answers, explaining why a certain answer reflects a common misconception or why a certain answer is correct.

Peer Instruction empowers students to create their own ideas, defend their own thoughts, and construct meaning with their peers. It motivates students and incites curiosity in the classroom. Peer Instruction allows students to experience the collaborating aspect of finding answers. Students at different levels of engagement benefit from science classrooms that use Peer Instruction. Peer Instruction works in a variety of disciplines, and, if implemented correctly, seems to improve conceptual knowledge, not just factual knowledge. Peer Instruction belongs in the tool box of every educator precisely because it is empowering and effective.

A participatory and engaged classroom

LEARNING WITH BODY IN MIND

"[A]s embodied, imaginative creatures, we never were separated or divorced from reality in the first place. What has always made science possible is our embodiment, not our transcendence of it, and our imagination, not our avoidance of it." — *George Lakoff*

"... don't underestimate the importance of body language" — *Ursula the witch in Disney's Little Mermaid*

Reimer, P., Mehta, R., & Mishra, P. (2019) Learning science with the body in mind. *iWonder : Rediscovering School Science, (5).* p. 24-28.

Consider explaining to someone the Earth's orbit around the Sun with your hands tied behind your back. Many people might find it challenging to talk about planetary motion without moving their hands or arms in an elliptical gesture. Many may even feel inadequate in explaining day-to-day experiences without using their bodies to express. This is because our understanding of the world is not exclusively encoded through language; our gestures and movements are also connected to the ways we think and learn.

We often think and understand the world using our bodies. Our senses and movement shape how we form and process knowledge. In understanding the world, it is quite common to

Our understanding of the world is not exclusively encoded through language; our gestures and movements are also connected to the ways we think and learn.

combine language and gestures to communicate meaningful concepts. Particularly when dealing with abstract concepts, we tend to base them in some concrete physical sensation. This is called embodied learning.

Scientists who study embodied learning are interested in the ways our bodies help us to learn and understand. They consider our body's sensory and motor activity as an essential part of learning, knowing, and meaning-making. They also encourage educators to reconsider the role of senses and physical movement in learning and design pedagogy that embraces embodiment.

What is Embodied Design for Learning?

Embodied design for learning considers how students can develop meaningful ideas using natural and intuitive movements. Students might engage in gestures or motion using their own natural movements and environments as learning resources. Then, combining physical movement with reflective language, students could explore and communicate their understanding of fundamental but abstract scientific concepts such as force, inertia, or motion. Thus, the physical actions they engage in become the pathway toward a deeper understanding of abstract concepts.

Embodied learning is not a stylistic preference like visual or kinesthetic learning styles. Embodied design principles concern with immersive, full-body thinking and knowing where physical movement is a natural part of learning. Additionally, embodied design for learning challenges traditional and common practices that privilege certain ways of knowing over others. In fact, scientific knowledge that resides in the head may not be superior to that which is embodied through action, often represented through muscle memory, instinct, visual mapping, or ways of moving.

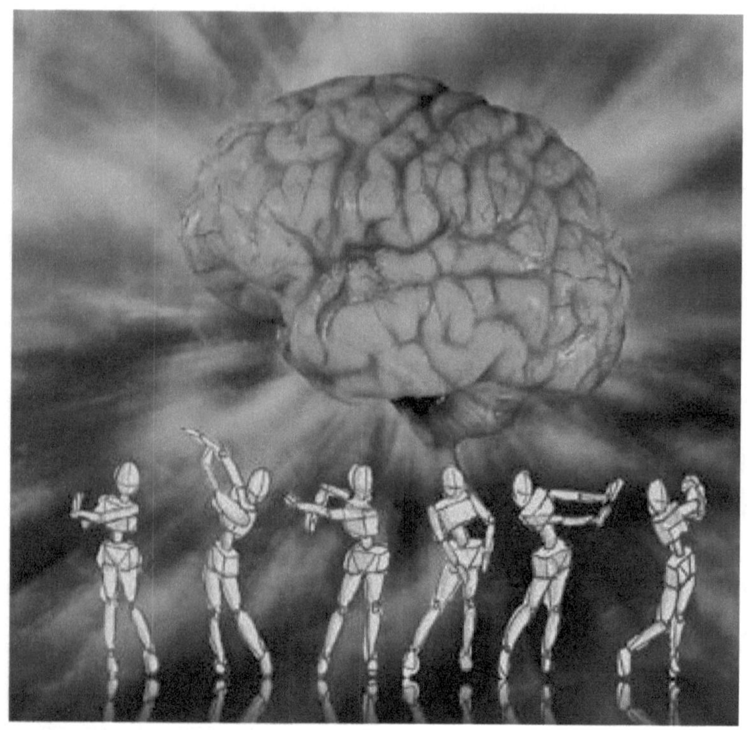

Physical actions they engage in become the pathway toward a deeper understanding of abstract concepts.

Through embodied design, science learning can become a more inclusive, participatory, and humanizing process, both in what is learned and how it is learned.

What Does the Research Say?

Roni Zohar and colleagues utilized an embodied design to support students' understanding of high-school concepts in physics. Their pedagogical design engaged students in physical experiences in which they coordinated movements to enact concepts. Then, they reflected on their experiences through the lens of the relevant physics ideas. We share examples of embodied design through two of their case studies.\

Case Study #1: In a dance studio, Zohar facilitated students as they learned about two concepts related to balance: area of support and center of mass. In four 90-minute lessons, students worked collaboratively to explore different bodily positions that were either steady or unsteady. Students learned that positions with larger area of support were more stable. Students also explored the concept of center of mass by considering where their center of mass was located when assuming different bodily positions, such as when raising their arms or leaning forward.

In the last lessons, students worked in pairs, with one student demonstrating a position while the other determined the area of support and the center of mass. The final project required pairs of students to demonstrate a sequence of three balanced positions while explaining their conceptual understanding using the terminology from the lessons as well as physical movements and gestures.

Case Study #2: A series of lessons focused on the difference between linear and angular velocity. Linear velocity was taught in a typical classroom setting, and angular velocity was taught in the dance studio. Students began by exploring circular movements with one body part at a time, such as their heads, hands, hips or feet. Each student explored circular movement at their own speed and direction.

The next activity required them to stand alongside holding hands and move around a bottle placed on the ground next to the person at the end of the line. Their movement as a row around the bottle at the center formed a circle of which they were the radius. To keep the angular velocity constant, the students had to negotiate their movements several times before reaching a consensus. To help the students understand the circular movement, the instructor also mimicked their movement using her arm, anchoring her elbow to the ground as a center and rotating her arm in a counter-clockwise motion. This movement represented the larger physical activity and further developed the concept of angular velocity.

In their analysis of both the case studies, Zohar and colleagues found that students' final projects demonstrated a deep understanding of the physics concepts taught. Students

Relating linear and angular motion. Photo of dancers overlaid with equations comparing linear and angular motion.

enacted their conceptual understanding of balance and velocity by using gestures and physical representations of the movement activities involved in the lessons. Making use of a variety of forms, such as music, art, dance, and video, their projects reflected their embodied experiences of the pedagogical design.

How Can Teachers Use Embodied Design to Support Science Learning?

Guide 1: Begin with Physical Experiences
Physical experiences have the potential to engage students and are often the hallmark of good teaching practice. Through the embodied design lens, physical experiences can take on new significance. These experiences serve as a starting point for the development of sophisticated conceptual understanding.

As Zohar and colleagues suggested, these bodily experiences "act as a resource enabling [learners] to relate complex (often abstract) ideas in physics to [their] everyday experiences" (2017, p. 68). Additionally, physical interactions bring value to various ways of thinking and knowing and can engage learners who may lack prior formalized knowledge. This important shift in the science classroom gives students access to concepts that might otherwise seem "out there".

One way teachers can put these ideas into practice is through the use of gestures. Studies have shown that when students gesture with their hands while they communicate, these movements help to form ideas for which they may not yet have language. As they learn, students seek to reconcile differences between their physical movements and oral expression.

Teachers can help to shape students' learning experiences

Understanding Newton's Laws through physical activity. Silhouette of children playing tug-of-war superimposed on text of Newton's notes on the Laws of Motion.

---

through the use of their own gestures as well. For instance, Zohar and colleagues also found that students imitated the instructor's gesture when asked to answer an oral question about the velocity lesson.

How to Implement: Take some time in instructional planning to explore possible physical actions that can lead to science conceptual understanding. You can start by analyzing your own physical movement and gestures as you explain the concepts. During planning, perhaps with your colleagues, consider the concepts that students can enact through movement. For example, imagine ways students might position their arms to simulate aircraft wings when studying flight mechanics. In classroom lessons, take advantage of students' natural movements to explore the dynamics of ecosystems. When introducing abstract or counter-intuitive ideas or concepts, be prepared to invite students to explore and move in new ways through scaffolded support.

Guide 2: Engage Learners in Reflection

A key component of embodied design for learning is to "guide learners from immersive action to structured reflection" (Abrahamson & Lindgren, 2014, p. 359). Reflective conversations can help students use language to explain their movements and, in process, make sense of the concepts. Through classroom dialogue, teachers can anchor new learning in students' lived experiences and encourage the development of meaning and collective understanding.

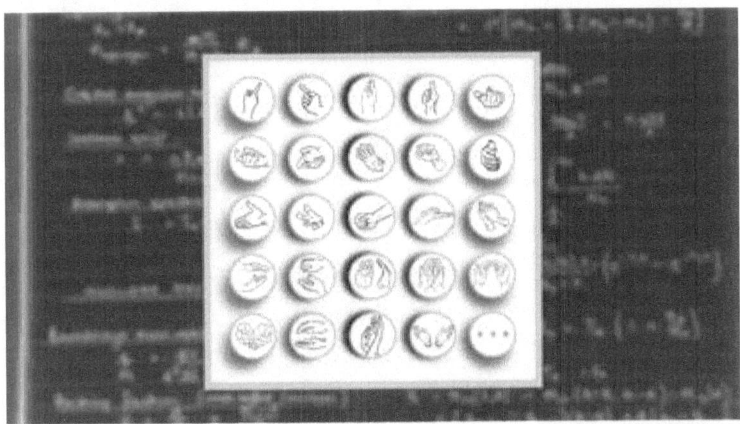

Take some time in instructional planning to explore possible physical actions that can lead to science conceptual understanding.

Creative reflective tools such as journals, video blogs, and interviews can also bring value and empathy to classroom interactions.

How to Implement: Provide opportunities for students to engage in reflective conversation about their physical experiences. Ask yourself: How did the experience challenge their assumptions and prior thinking? How did the physical movements disrupt their existing patterns of thinking about the concepts? What new ideas or questions do they have now? Teachers develop a humanizing learning environment when classroom conversations focus on highlighting students' strategies and insights.

Guide 3: Encourage Multimodal Student Projects
Drawing on the strengths of project-based learning, multimodal projects provide students opportunities to draw on a variety of resources to develop understanding and communicate meaning. Collaborative student projects move beyond written word responses and can take the form of art, video, dance, oral presentation, or short film. These projects give students increased participation and allow for the inclusion of bodily movement in creating and sharing

Provide opportunities for students to engage in reflective conversation about their physical experiences.

meaning. Projects that leverage more tools—such as brain, body, and environment—have the added benefit of enhancing creativity and expression.

How to Implement: Give students opportunities to incorporate social, cultural, and personal elements into science projects. While poster boards and three-dimensional models are well-suited to scientific displays, physical movement is less common. Consistent with embodied design, teachers might encourage a movement routine to demonstrate how organisms interact within their environment. Perhaps students might use their hands to illustrate the push and pull of magnetic forces, how electricity travels along a circuit, or ideas related to the transfer of energy.

Collaborative student projects move beyond written word responses and can take the form of art, video, dance, oral presentation, or short film.

Conclusion

Embodied design for learning presents several unique challenges to the ways we conceptualize thinking and learning. For science teachers, embodied design highlights the role of physical movement in how our students interact with important scientific ideas and processes. Embodied design presents opportunities for us to rethink our science teaching practices. In many ways, it offers us a pedagogy that recasts learning as a more complete, complex and human activity.

REFERENCES

Chapter 1: Why care about beauty

Flannery, M. C. (1991). Science and aesthetics: A partnership for science education. *Science Education, 75*(5), 577-593. doi:10.1002/sce.3730750507.

Girod, M. & Wong, D. (2001). An aesthetic (Deweyan) perspective on science learning: Case studies of three fourth graders. The *Elementary School Journal, 102*(3), 199-224.

Girod, M., Rau, C., & Schepige, A. (2003). Appreciating the beauty of science ideas: Teaching for aesthetic understanding. *Science Education, 87*(4), 574-587.

Girod, M., Twyman, T., & Wojcikiewicz, S. (2010). Teaching and learning science for transformative, aesthetic experience. *Journal of Science Teacher Education, 21*, 801-824.

Chapter 2: Bringing in social justice

Calabrese Barton, A., Birmingham, D., Sato, T., Tan, E., & Calabrese Barton, S. (2013). Youth As Community Science Experts in Green Energy Technology. *After school Matters*. Retrieved from http://eric.ed.gov/?id=EJ1016811

Dimick, A. S. (2012). Student empowerment in an environmental science classroom: Toward a framework for social justice science education. *Science Education, 96*(6), 990–1012.

Langhout, R. D., Collins, C., & Ellison, E. R. (2014). Examining Relational Empowerment for Elementary School Students in a yPAR Program. *American Journal of Community Psychology, 53*(3-4), 369–381. http://doi.org/10.1007/s10464-013-9617-z

Mueller, M., Tippins, D., Bryan, L. (2012). The Future of Citizen Science. *Democracy & Education. 20*(1), 1-12.

Chapter 3: Seeing with multiple metaphors

Feynman, R. P., Leighton, R. B., & Sands, M. (2015). *The Feynman lectures on physics, Vol. I: The new millennium edition: mainly mechanics, radiation, and heat (Vol. 1)*. Basic Books.

Lancor, R. (2014). Using metaphor theory to examine conceptions of energy in biology, chemistry, and physics. *Science & Education, 23*(6), 1245-1267.

Lancor, R. A. (2013). The many metaphors of energy: Using analogies as a formative assessment tool. *Journal of College Science Teaching, 42*(3), 38-45.

Ledford, H. (2015). How to solve the world's biggest problems. *Nature,* 525, 308-311.

Wieman, C. (2007). Why not try a scientific approach to science education? *Change: The Magazine of Higher Learning, 39*(5), 9-15.

Chapter 4: Students as teachers

Crouch, C. H., & Mazur, E. (2001). Peer instruction: Ten years of experience and results. *American Journal of Physics (69)*970. https://doi.org/10.1119/1.1374249

Judson, E. & Sawada, a.D. (2002). Learning from Past and Present: Electronic Response Systems in College Lecture Halls. *Journal of Computers in Mathematics and Science Teaching, 21*(2), 167-181. Norfolk, VA.

Mazur, E. (1997). *Peer Instruction: A User's Manual.* Prentice–Hall, Upper Saddle River, NJ.

Abrahamson, D., & Lindgren, R. (2014). Embodiment and embodied design. In K. Sawyer (Ed.), *The Cambridge Handbook of the Learning Sciences* (2nd ed., pp. 358-376). Cambridge, UK: CUP.

Zohar R., Bagno, E., & Eylon, B. (2015). Dance and movement as means to promote physics learning. In *Proceedings of the 7th International Conference on Education and New Learning Technologies* (pp 6881-6885) Barcelona: EDULEARN15.

Zohar, R., Bagno, E., Eylon, B., & Abrahamson, D. (2017). Motor skills, creativity, and cognition in learning physics concepts. *Functional Neurology, Rehabilitation, and Ergonomics, 7*(3), 67–76.

CONTRIBUTORS

Nicole Bowers, a former k-12 teacher, is currently doctoral student in the *Mary Lou Fulton Teachers College* at *Arizona State University*. Her research interests include environmental and sustainability education, particularly community and place-based pedagogies.

Kevin Close dabbles in education and design with the enthusiasm of a kindergärtner, the efficiency of a freshman, and with dreams of getting a PhD. When not writing, Kevin devotes his time to redesigning and re-imagining the measurement of student achievement. He takes a practical approach to designing 'smart' standardized tests that adapt for diverse students.

Day Greenberg is a post-doctoral researcher at *University of Michigan*, studying learning and development experiences of youth engaged in out-of-school STEM programs. She specializes in new critical participatory and qualitative research approaches, youth-adult partnership initiatives, and the community-centered design of learning spaces, programs, and curricula.

Dr. J. Bryan Henderson Assistant Professor of *Learning Sciences* at *Arizona State University*. He is interested is in the utilization of educational technology to facilitate critical, peer-to-peer science learning. He is a *National Academy of Education/Spencer Fellow*, editorial board member for the *Journal of Research in Science Teaching*, and program chair of the *Science Teaching and Learning* SIG of the *American Educational Research Association*.

Sarah Keenan-Lechel received her PhD in *Educational Psychology Educational Technology* from *Michigan State University*. Her research focuses on equitable science education, examining the ways in which we can invite youth into the research process as a way to disrupt multiple levels of power dynamics.

Mashood K.K. is Reader at *Homi Bhabha Centre for Science Education, Tata Institute of Fundamental Research*, Mumbai, India. His interest is in science education research, particularly physics. Mashood can be reached at mashood@hbcse.tifr.res.in

Dr. Rohit Mehta is assistant professor of *Secondary Curriculum with Instructional Technology* in the *Department of Curriculum and Instruction* at *California State University, Fresno*. He conducts inquiries on the intersections of creativity, epistemology, and social power. He often writes about technology in education and what it means for local and global literacies.

Punya Mishra is an educator, researcher, designer, and professional dilettante interested in life, the universe and everything. He is particularly keen on shoehorning Douglas Adams' book titles into his bio statement. You can find him at @punyamishra and punyamishra.com

Paul Reimer is a doctoral candidate in *Educational Psychology and Educational Technology* at *Michigan State University*. His research is focused on the learning sciences with an emphasis on playful participation in mathematics and science education. He shares his work online at @reimerpaul and paulnreimer.com.

9 781087 288581